PERFECT PREDATORS

GREAT WHITE SHARKS

ON THE HUNT

Robin Johnson

A Crabtree Forest Book

Crabtree Publishing
crabtreebooks.com

Notes for Readers and Class Discussion

This book is designed to teach readers about core subject areas, build curiosity, and inspire further investigation. Readers are encouraged to build upon what they already know about the subject and engage in topics they want to learn more about. Here are a few guiding questions to prompt reflection and discussion. Possible answers appear in red.

Before Reading

Read the title and look at the table of contents. What do you already know about predators and great white sharks?

- *I know that predators are animals that hunt and eat other animals.*
- *I know that great white sharks are large predators that live in oceans all over the world.*

What would you like to learn about great white sharks?

- *I would like to learn about the body parts that make great white sharks such deadly hunters.*
- *I would like to learn how great white sharks compare to other ocean predators.*

During Reading

Pause after reading each page or chapter. What questions do you have about what you read? What are you curious to know?

- *I wonder why the megalodon became extinct if it was such a large and fierce predator. When did the last megalodon die?*
- *I wonder why great white sharks usually live alone. I wonder if this is similar to or different from other top predators.*

Make connections with what you are reading. How is this information similar to something you already know?

- *I know that the gray wolf is an apex predator where I live. It helps control the populations of prey animals such as deer.*
- *I know that climate change is affecting the habitats of many animals because rising temperatures change the type of food that is available.*

After Reading

Recall key details about the book. What was the author trying to teach readers?

- *The author was trying to teach readers that a great white shark's physical abilities and sharp senses make it an especially deadly hunter.*
- *The author was trying to teach readers that humans are the biggest threat to great white sharks.*

How did the images and captions help you understand more?

- *The diagrams and labels helped show me how a great white shark's different body parts help it find and attack prey.*
- *The photographs showed me how well a great white shark can blend into its surroundings.*

CONTENTS

CHOMP! TEAR! KILL!

What's so great about the great white shark? Is it the animal's massive body, powerful jaws, and rows of great white teeth? Is it the incredible speed at which this fearsome fish tears through the ocean? Or is it this perfect **predator**'s unmatched ability to hunt and kill its **prey**? The answer is all the above!

Named for their white undersides, great white sharks are also called white sharks, white pointers, great whites, and—most chillingly—white death. Most scientists use the name "white shark."

WHAT IS A PERFECT PREDATOR?

There are many meat-eating animals lurking in the depths of the ocean and in other habitats around the world, but only some are perfect predators. These predators are smart, fast, and often more powerful than any other animals. They use their jaws, claws, and other body parts as deadly weapons, surprising prey with their speed, strength, **stealth**, and fierce attacks.

CHAPTER 1

KILLING

The great white shark is a killing machine built for speed and sharp and sly in its attack. The size and shape of this animal—not to mention its powerful jaws and sharp teeth—help make it a perfect predator.

MACHINES

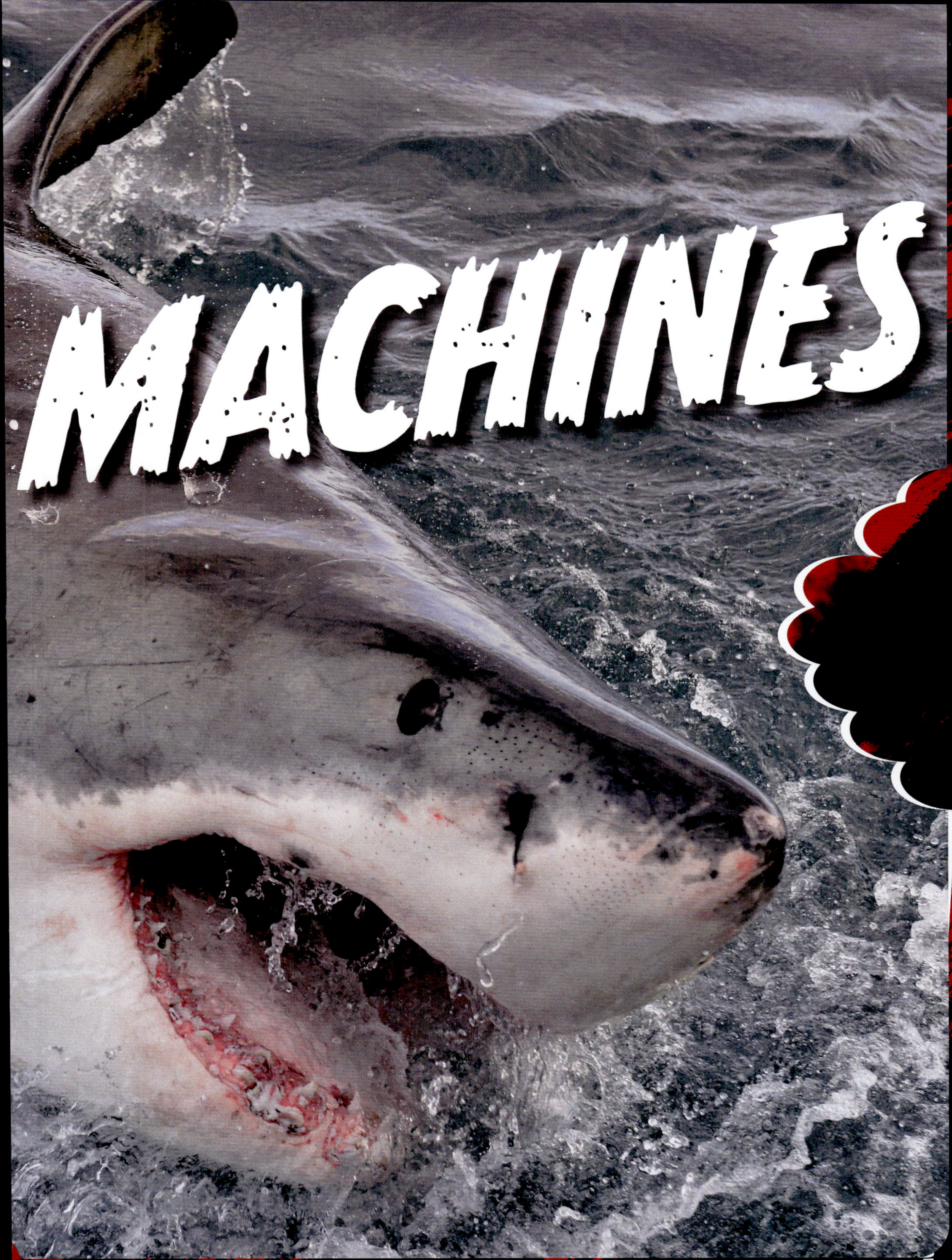

SIZE MATTERS

Ocean Predators

There are plenty of fish in the sea—and the great white shark is one of the biggest and fiercest of them all. It can reach up to 21 feet (6.4 m) long and weigh up to 4,500 pounds (2,040 kg). As with most shark species, female great white sharks are larger than males. But they are both massive beasts—and the largest fish that hunt and eat other animals.

The great white shark is the largest predatory fish in the world.

Bigger Fish

Great white sharks are huge, but they are not the biggest sharks in the sea. Whale sharks and basking sharks are both bigger and heavier than great white sharks. Luckily for many sea creatures, these sharks are filter feeders, not predators. They swim slowly through the ocean, scooping up small fish and tiny animals such as plankton with their giant mouths.

MEGA SHARK

The great white shark is not the deadliest shark to ever swim in the sea. An enormous, now-**extinct** shark called the megalodon terrorized the ocean millions of years ago. It reached up to 60 feet (18 m) in size, making it the largest fish in history.

This picture shows a megalodon's tooth (right) compared to the tooth of a great white shark. The name "megalodon" means "giant tooth."

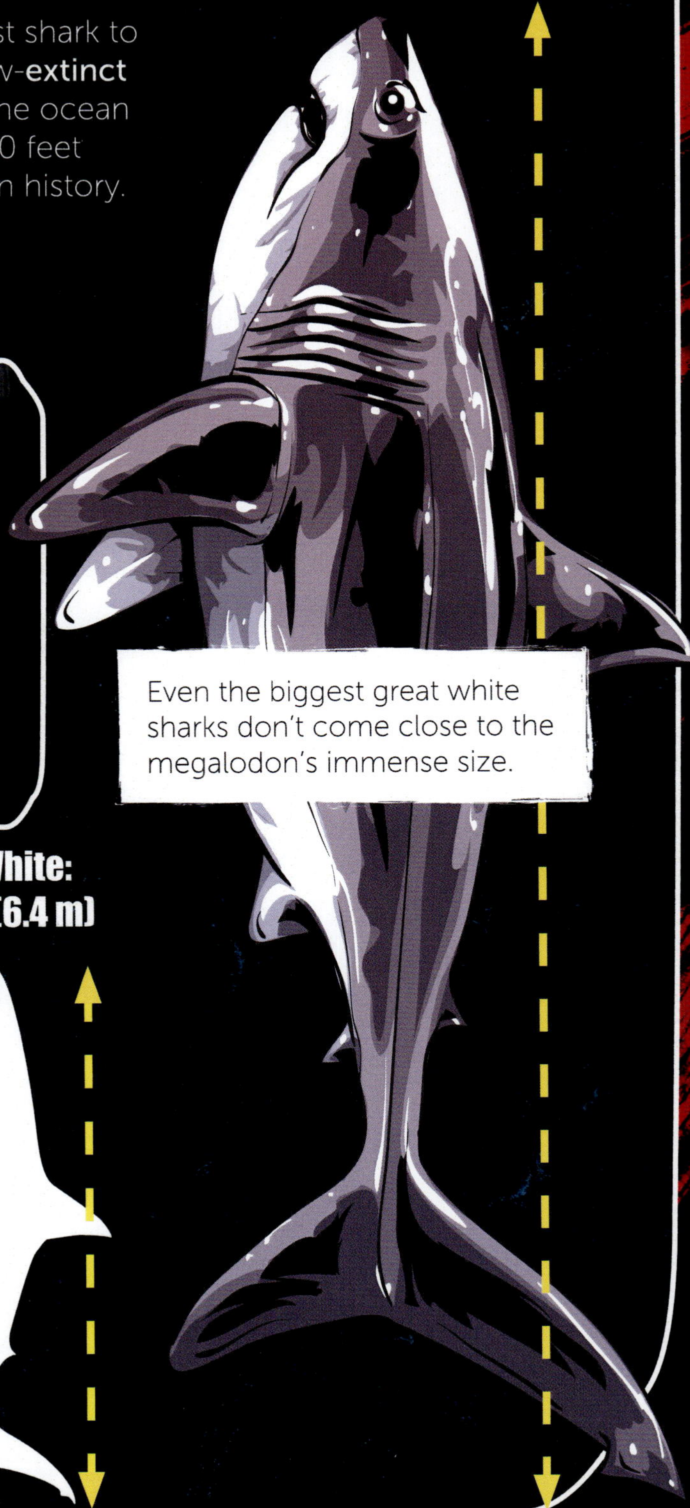

Even the biggest great white sharks don't come close to the megalodon's immense size.

SPEED DEMON

Built for Speed

Great white sharks are built like underwater weapons called torpedoes. This smooth, **streamlined** shape allows sharks to swim at speeds up to 35 miles per hour (56 kph), finding and destroying their targets.

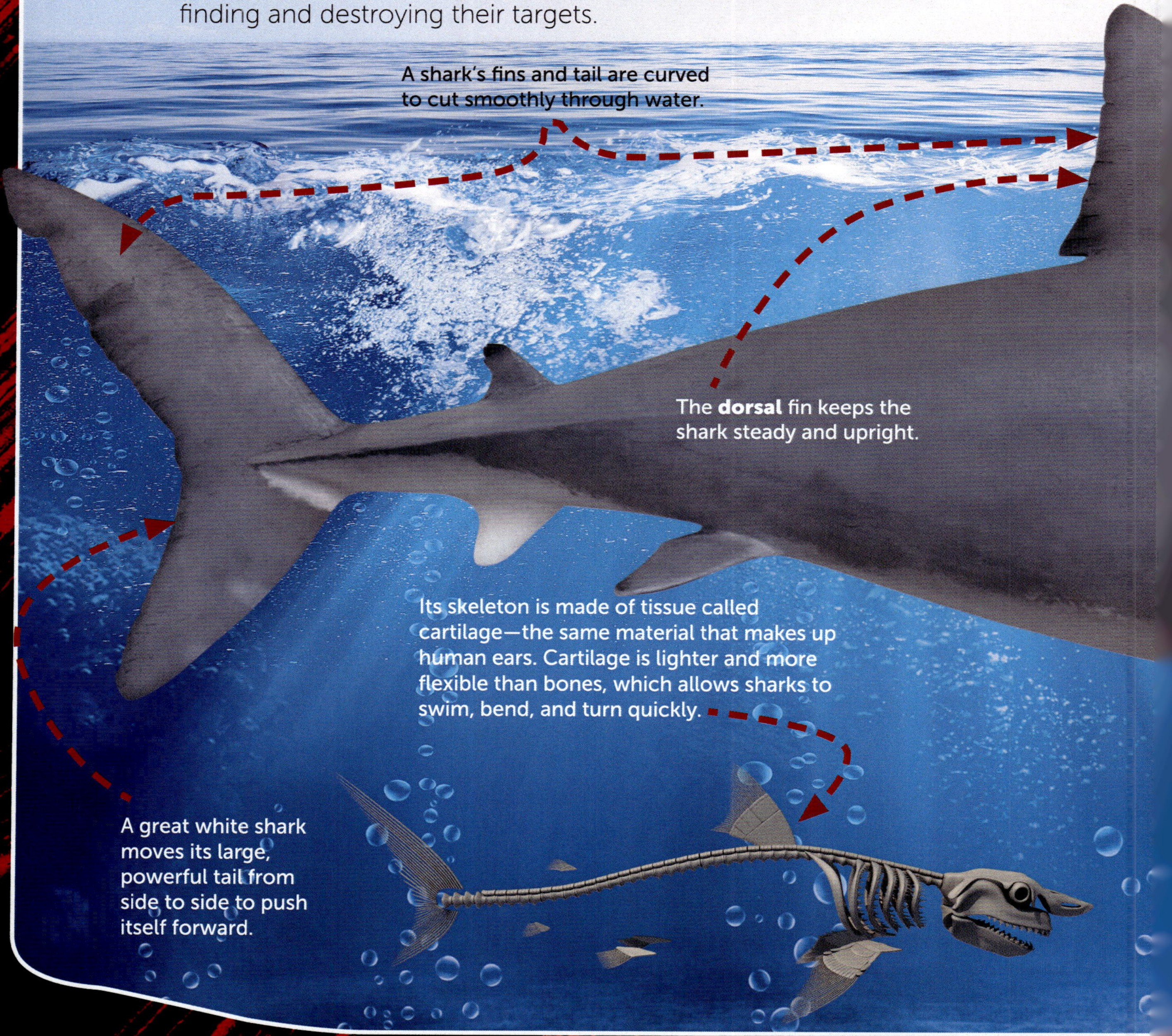

The great white shark's dorsal fin warns other animals—and people—that a deadly predator is approaching.

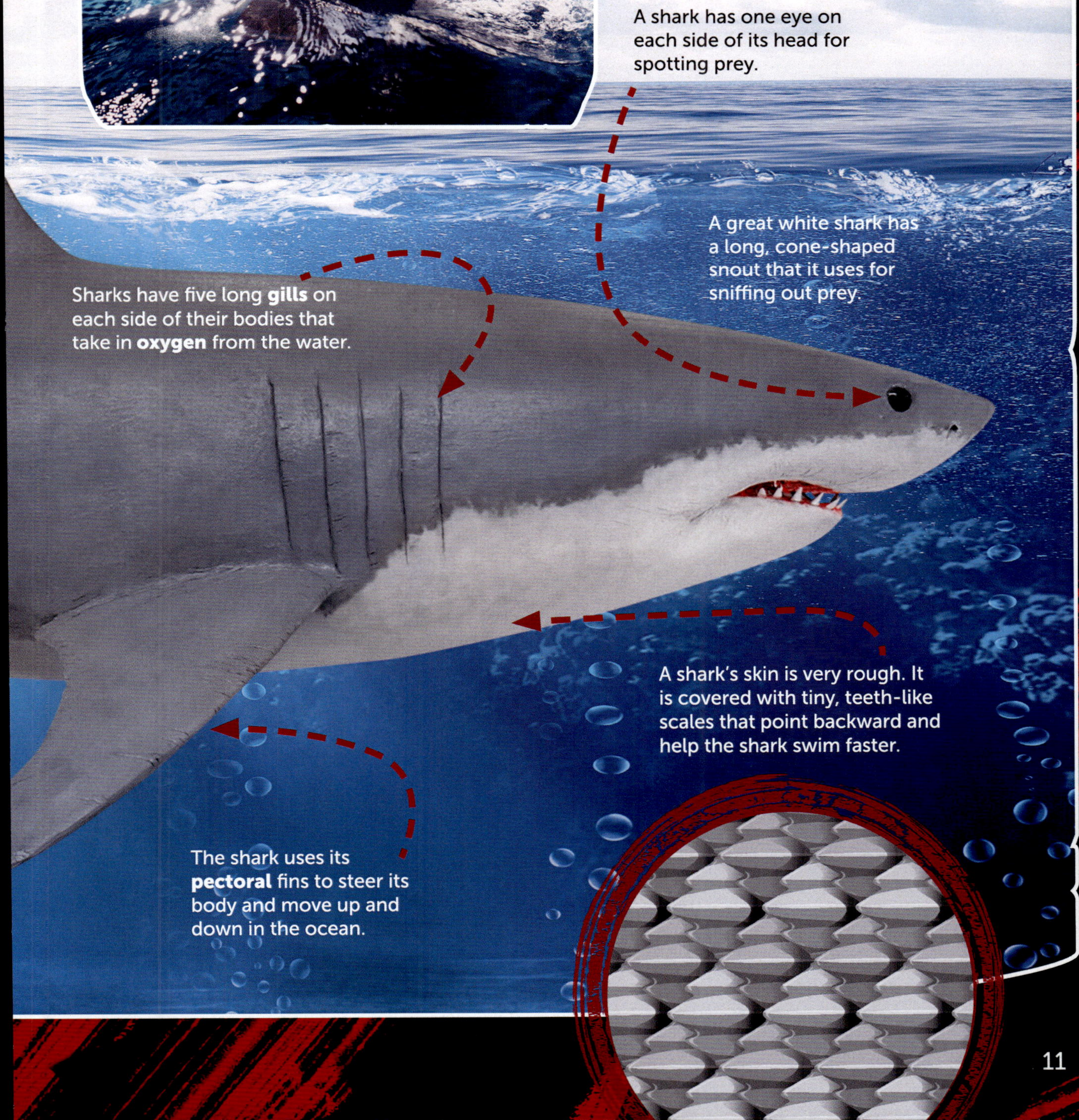

CHOMP!

Fear the Jaws

The great white shark's most famous feature is its jaws. They are powerful enough to tear through even the toughest flesh and shatter an animal's bones. Attached to the jaws are rows of large, triangular teeth—up to 300 in total. A great white shark constantly loses and replaces these teeth throughout its life.

Each great white shark tooth is about 2.5 inches (6 cm) long. It has a sharp, pointed tip and jagged, knife-like edges.

BITE FORCE ONE

The great white shark has one of the most powerful bites in the animal kingdom. Scientists used computer models to calculate the shark's bite force—from a safe distance. They discovered it has three times more force than a lion's bite!

PREDATOR POP QUIZ

Test your knowledge of great white shark bodies and choose the word that best finishes each sentence.

1 Great white sharks are the

biggest

smallest

deadliest

sharks in the sea.

2 Their skeletons are made of

bones

cartilage

pure muscle

3 A shark uses its

tail **fins** **gills**

to push itself through water and its

tail **fins** **gills**

to steer and balance its body.

4 A great white shark has about

100

200

300

razor-sharp teeth in its mouth.

5 It uses its large, cone-shaped snout to

breathe

smell prey

dig for food

Answers: 1. deadliest 2. cartilage 3. tail, fins 4. 300 5. smell prey

GREAT WHITE SHARK

Great white sharks and tiger sharks are both massive, aggressive predators—and the two shark species most likely to attack humans.

A great white shark has a pointed snout, while a tiger shark has a wide snout. They both have a keen sense of smell, however.

The great white shark has bigger, more powerful jaws than the tiger shark, so it can attack and eat larger prey animals.

Despite the great white shark's name, only its underside is white. The rest of its body is gray.

VS TIGER SHARK

Weighing up to 2,000 pounds (900 kg), the tiger shark is the second-largest predatory shark. Only the great white shark is bigger, stronger, and deadlier.

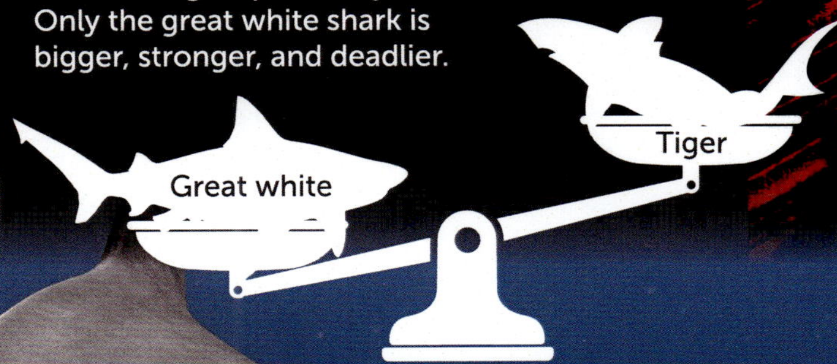

Great white shark tooth

Tiger shark tooth

Both shark species have sharp, saw-edged teeth that can cut through bones and turtle shells. The tiger shark has only 48 teeth, however, and they are smaller and more slanted than great white teeth.

Tiger sharks are greenish blue to gray, with white undersides.

Tiger sharks are **opportunistic** feeders that will eat just about anything they find, including fish, birds, and sea turtles. They also eat garbage such as tin cans and tires, and sometimes, people that they find in shallow areas of the ocean.

CHAPTER 2

TOP OF THE

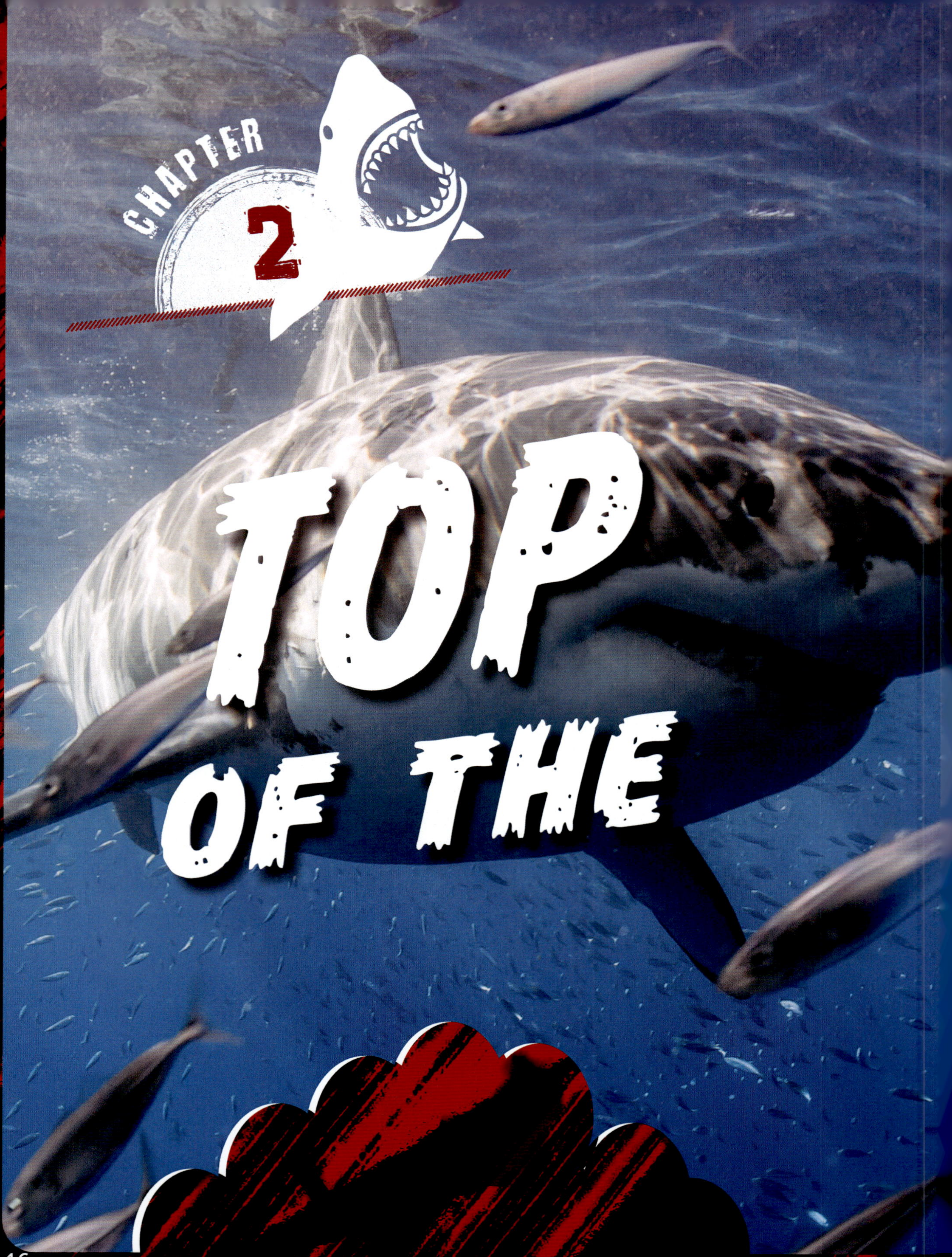

FOOD CHAIN

Great white sharks are **apex predators.** They are at the top of the marine **food chain** and have no natural predators. They are born ready to hunt and quickly grow into unstoppable killers.

NOWHERE TO HIDE

WHITE SHARK CAFÉ

There is a remote area in the middle of the Pacific Ocean where many great white sharks gather. They spend up to 100 days traveling from coastal habitats around the world to reach the area, which scientists have named "White Shark Café." Here the sharks grab a bite to eat. Prey animals are abundant deep below the surface there—and may stay in the area for several months.

Shark-Infested Waters

Great white sharks are found in every ocean on Earth. They live in **temperate** waters near coasts, where there are plenty of seals and other animals for them to hunt. The sharks swim mainly in waters less than 330 feet (100 m) deep, but they can dive to 4,000 feet (1,220 m) in search of prey.

On the Move

Some great white sharks also travel long distances in the open ocean. They **migrate** to different areas to find food, mate, or have their young. Scientists have tracked great white sharks swimming from South Africa to southern Australia and back again!

Most seals **breed** and raise their young in huge groups along coasts, which attracts great white sharks.

BORN TO KILL

Natural Killers

Great white sharks begin killing right from the start. Baby great white sharks grow in eggs that hatch inside the mother's **womb**. The sharks continue to grow and develop in the womb until they are born—if they make it out alive. Baby sharks eat other eggs in the womb. They also feed on their smaller brothers and sisters before they are born.

Mother's Nature

Shark babies, called pups, are born big and strong and ready to hunt. They are about 5 feet (1.5 m) long at birth. The mother shark does not care for her pups or protect them from danger. Instead, she tries to catch and eat them! The pups swim quickly away from their dangerous mother and will spend the rest of their lives on their own. Many do not survive their first year.

Great white shark pups are on their own from the moment they are born.

Hunters and Hunted

Young great white sharks hunt fish, rays, and small sharks. They are also hunted by many animals, including other great white sharks. As the pups grow bigger and more powerful, they face fewer natural threats. Over time, the sharks take their place at the top of the food chain.

BITE TO EAT

Shark Food

Great white sharks hunt and eat a variety of marine animals. Their favorite food is seals, but they also eat sea lions, dolphins, small whales, porpoises, otters, and sea turtles. Many of these animals have a thick layer of fat called blubber that provides sharks with the energy they need to survive. Sharks usually don't waste their time and energy eating small animals.

Feeding Frenzy

Great white sharks are also scavengers. These are animals that feed on dead animals. Great white sharks feed on dead whales or the remains of other large sharks they find floating in the ocean. If several sharks gather to feed on the same animal, sometimes it turns into a feeding frenzy. The sharks go wild and start biting everything in sight—including each other.

This great white shark has found a plump seal to eat.

OCEAN VIPS

Great white sharks are VIPs—very important predators—-in the ocean. Like all apex predators, they help control the **populations** of the animals they eat. They also keep species healthy by targeting sick or weak animals. Without great white sharks, there would be too many seals and other marine mammals, making the entire **ecosystem** unbalanced.

Great white sharks can leap out of the water, or breach, to catch prey.

A great white shark and some seabirds are feeding on the remains of a humpback whale.

CHAPTER 3

SHARK

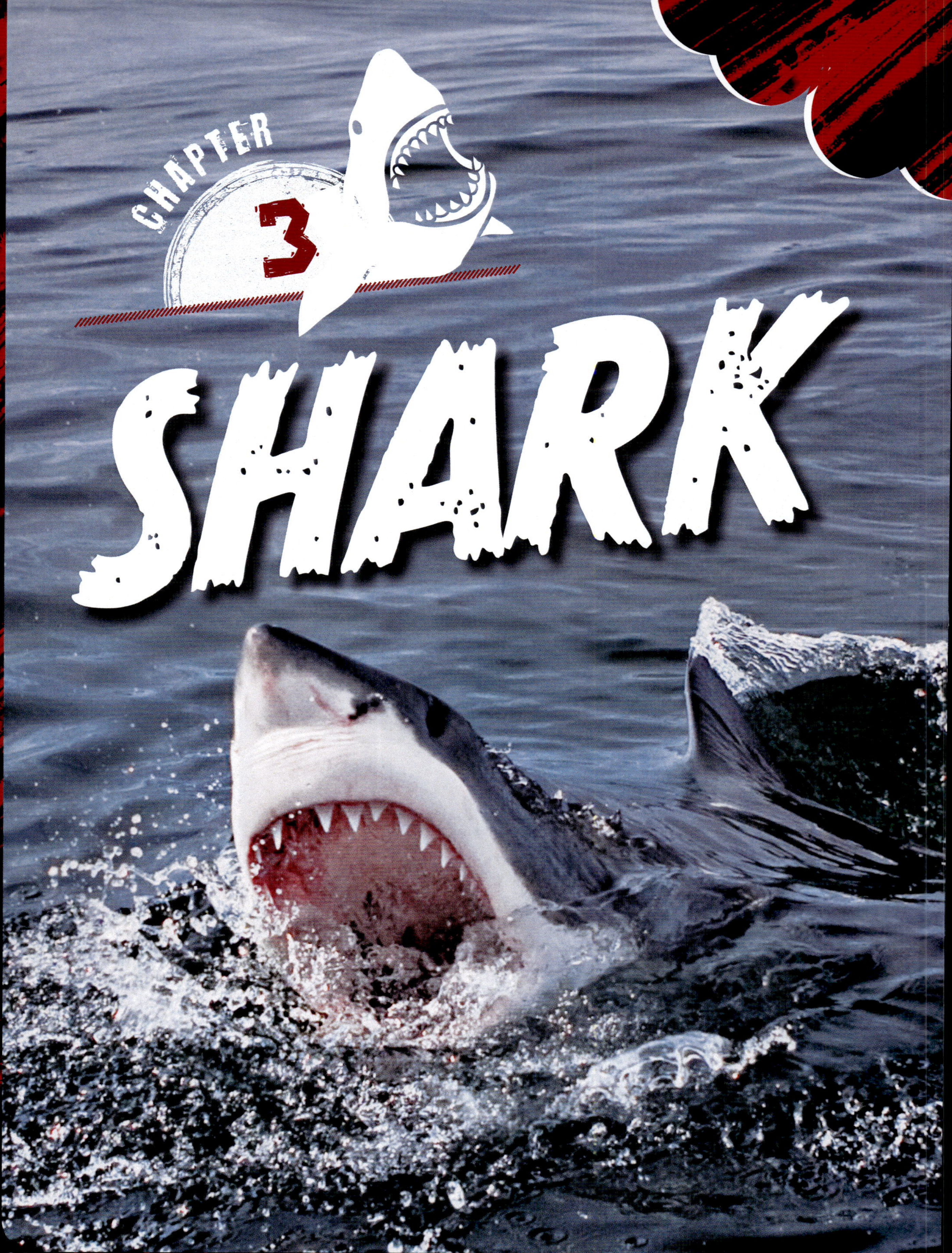

ATTACK

Great white sharks are ambush predators. That means they use the element of surprise to suddenly capture and kill their prey. They strike with speed, force, and **precision**. Few animals can survive a targeted attack by a great white shark.

SUPER SENSES

Senses for Searching

Great white sharks are well **adapted** for hunting in their habitats. They use all their senses to locate prey in dark, open ocean waters.

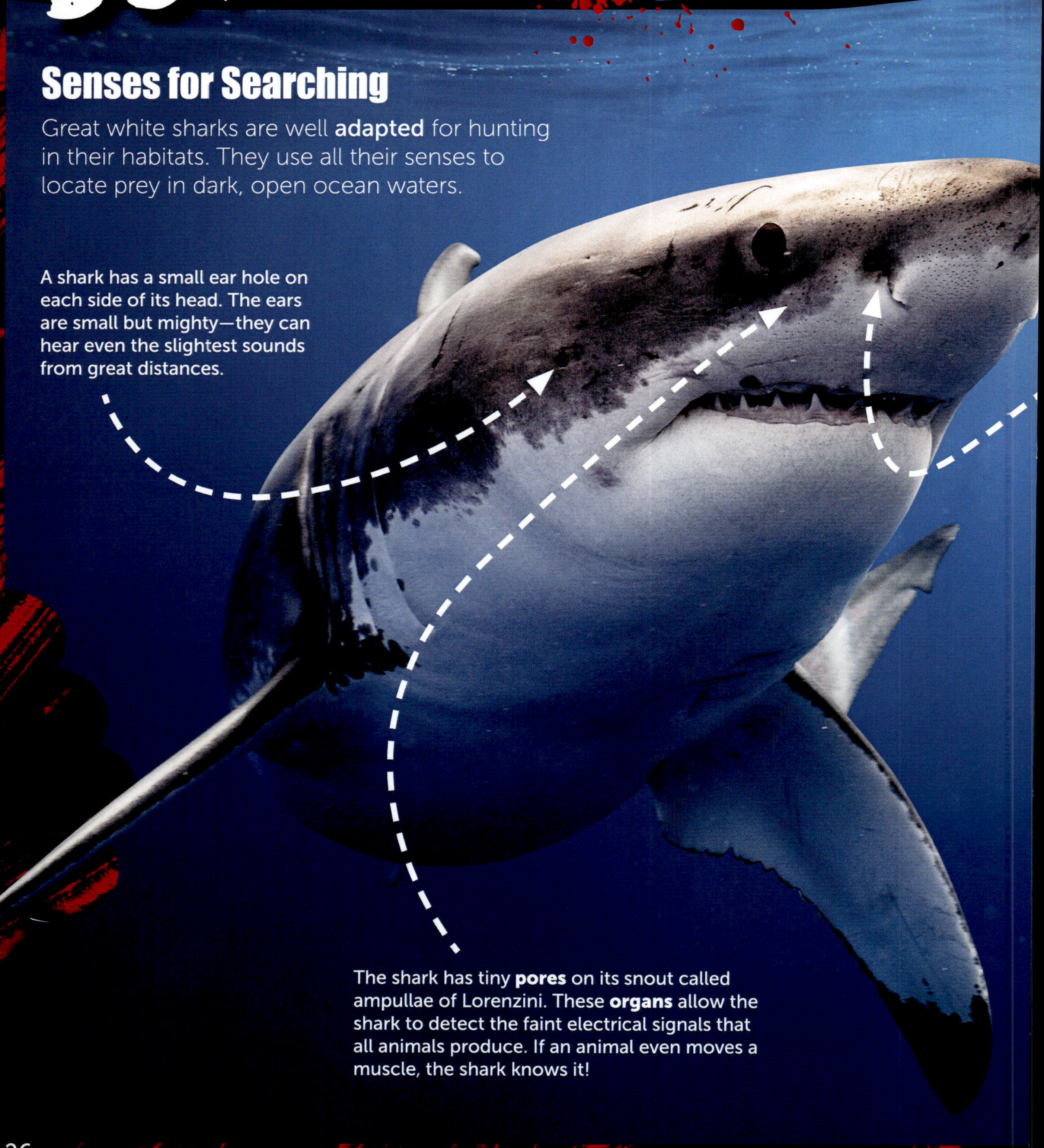

A shark has a small ear hole on each side of its head. The ears are small but mighty—they can hear even the slightest sounds from great distances.

The shark has tiny **pores** on its snout called ampullae of Lorenzini. These **organs** allow the shark to detect the faint electrical signals that all animals produce. If an animal even moves a muscle, the shark knows it!

A great white shark has a keen sense of touch. Like all fish, it has a **lateral** line that runs down the middle of its body from head to tail. This line allows the shark to sense even the smallest movements and pressure changes in the water.

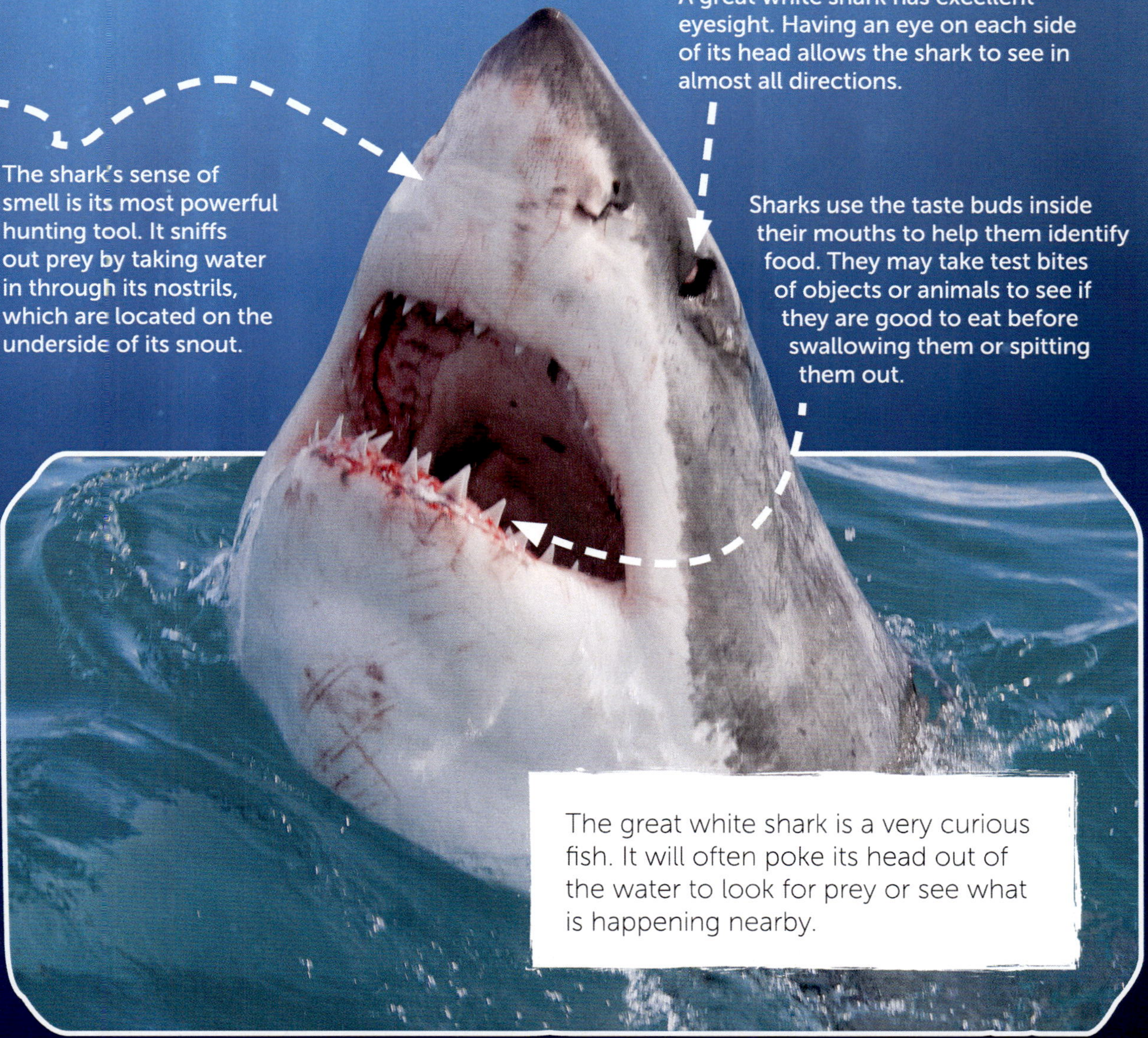

A great white shark has excellent eyesight. Having an eye on each side of its head allows the shark to see in almost all directions.

The shark's sense of smell is its most powerful hunting tool. It sniffs out prey by taking water in through its nostrils, which are located on the underside of its snout.

Sharks use the taste buds inside their mouths to help them identify food. They may take test bites of objects or animals to see if they are good to eat before swallowing them or spitting them out.

The great white shark is a very curious fish. It will often poke its head out of the water to look for prey or see what is happening nearby.

A HIDDEN THREAT

Hide and Seek

Like all predators, great white sharks must avoid being spotted by the animals they hunt. They usually search for prey alone, moving noiselessly through the ocean. They often hunt at dusk or dawn when sunlight is dim. Sharks see well even in low light, which gives them an advantage over many prey animals when they are hiding and seeking in the sea.

The great white shark's countershading helps it to blend in with its ocean environment.

COUNTERSHADING

To help them stay hidden in ocean waters, great white sharks have a type of **camouflage** called countershading. That means the upper side of a shark's body is darker than the lower side. When a prey animal swims above a shark, the shark's gray back blends in with the dark ocean waters below. When an animal swims below the shark, the shark's white underside blends in with the bright sunlit surface of the water above.

PREDATOR POP QUIZ

See if you can spot the sharks lurking in these pictures. They are especially hard to see in dark or murky ocean waters.

ATTACK!

On the Hunt

Great white sharks are smart, highly skilled hunters. They can detect if animals are weak, wounded, or struggling with other predators by the sounds and movements they make, even from great distances. The sharks target these animals because they take less energy to kill.

Charge from Below

A great white shark usually hunts near the surface of the ocean. It tracks its prey and then moves in for the kill. The shark swims in circles beneath the unsuspecting animal, slowly getting closer and closer to it. Then the shark charges toward its prey with a sudden burst of speed.

Biting and Breaching

The shark hurls its huge body upward, crashing into its prey and biting it at the same time. The shark often bursts right out of the ocean in a spectacular leap called a breach. Then it lands back in the water with the shocked animal firmly in its jaws.

Sometimes the speed and force of an attack hurls the shark and its prey right out of the water!

DEADLY BITE

Blood in the water signals a fresh kill by a great white shark.

Bite and Wait

One bite from a great white shark is usually all it takes to kill or seriously injure an animal. The shark swims a short distance away and waits for the wounded animal to die or become too weak to fight. Then the shark returns and eats its fresh kill. Sharks have plenty of sharp teeth but do not chew their food. Instead, they tear off bite-sized pieces of meat and swallow them whole.

Great white sharks are sometimes called "swimming noses" because of their excellent sense of smell.

PREDATOR POP QUIZ

Here are some statements about great white sharks for you to chew on. See if you can separate fact from fiction!

1 A great white shark could smell a single drop of blood in an Olympic-size swimming pool.

2 A group of great white sharks is called a fright.

3

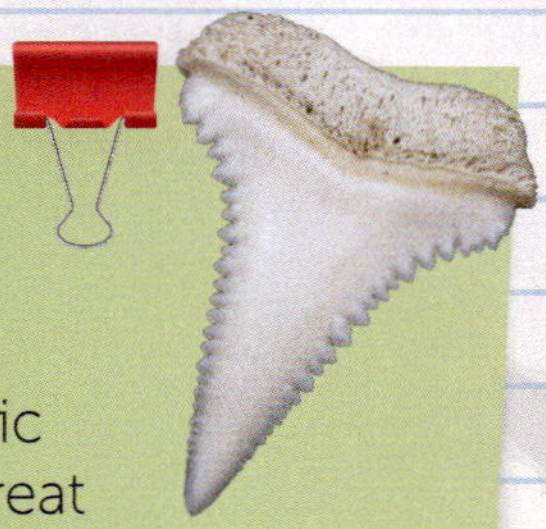

The scientific name for great white shark means "ragged tooth."

4 Great white sharks can survive up to six months between feedings.

ANSWERS

1. **FACT!** A shark's nose knows the way to prey!
2. **FICTION** The name for a group of sharks—and other fish—is a school or shoal.
3. **FACT!** A shark's teeth are ragged, jagged, and ready to chomp!
4. **FICTION** They can survive up to two months between meals, but they get pretty hangry!

GREAT WHITE SHARK

Great white sharks have no natural predators, but orcas may attack and kill these sharks from time to time. Orcas are huge, fierce ocean mammals that are also known as killer whales.

Orcas are much bigger and heavier than great white sharks. They can grow up to 32 feet (9.7 m) long and weigh up to 16,000 pounds (7,200 kg).

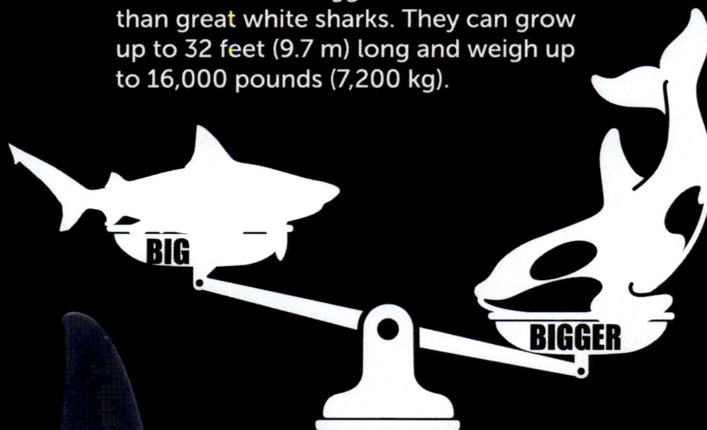

Great white sharks usually live alone, but they may hunt in pairs or small groups from time to time.

Sharks rely on their well-developed senses to find food.

Sharks do not like to share their food.

A great white shark is a fast and furious fish.

Both animals have streamlined bodies and can swim quickly through water, but the shark has faster bursts of speed for attacking prey.

VS ORCA

Orcas are social animals. They live and hunt in large family groups called pods.

Orcas use **echolocation** to find prey in dark ocean waters. That means they make clicking sounds and listen for the echoes that bounce off objects. This tells orcas the size, shape, speed, and direction of the objects.

Orcas are black on top and white on the bottom. Like great white sharks, they rely on countershading to stay hidden from prey.

An orca is neither a whale nor a fish. It is the largest member of the dolphin family.

Orcas work together to surround and capture seals, sea lions, fish, and other prey. They even share food with one another.

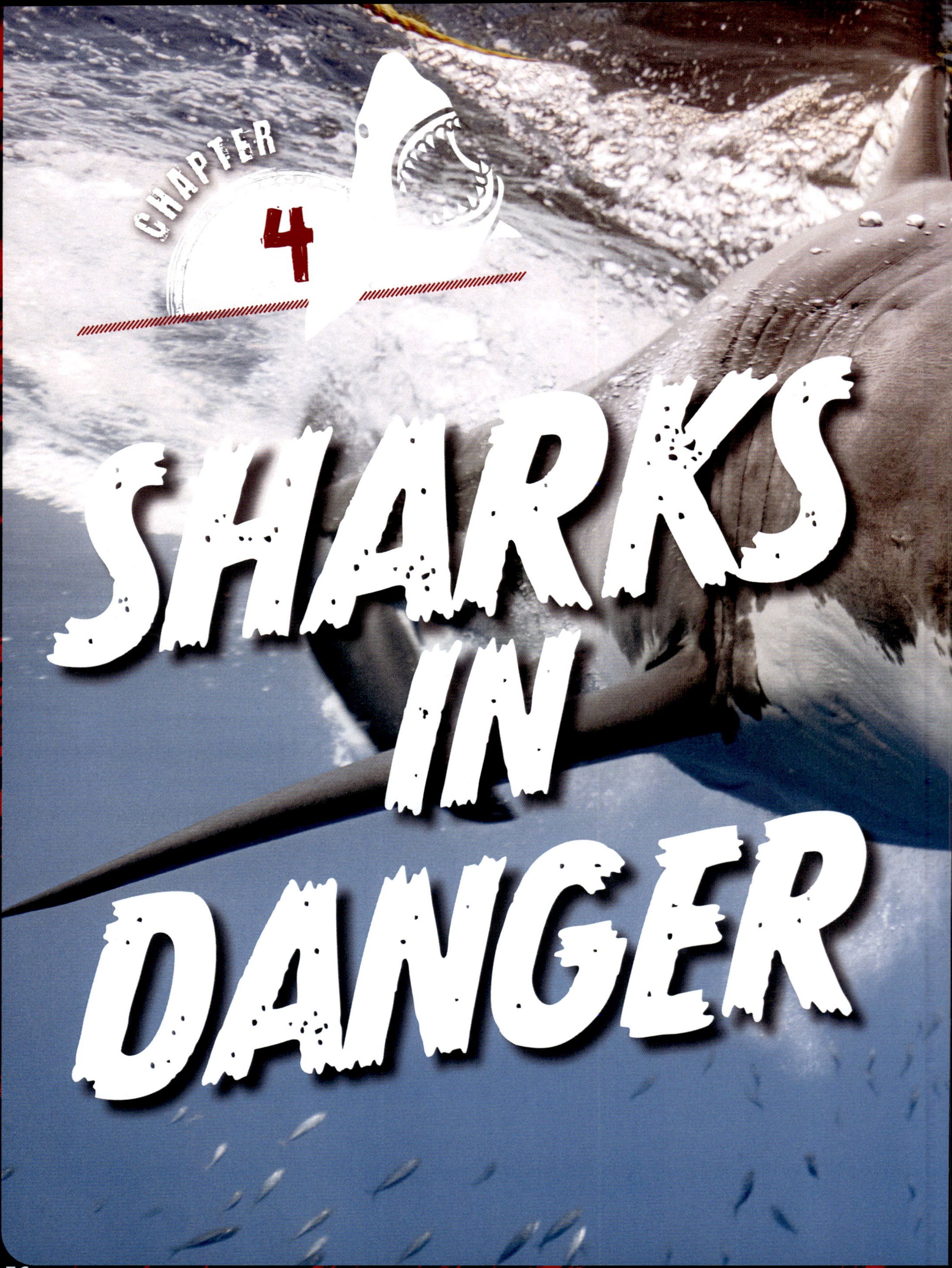

CHAPTER 4

SHARKS IN DANGER

Great white sharks rule the ocean and are too strong and swift for most animals to kill. The only real threat to great white sharks is humans. People around the world kill great white sharks, but others are working to save this species.

HUMAN THREAT

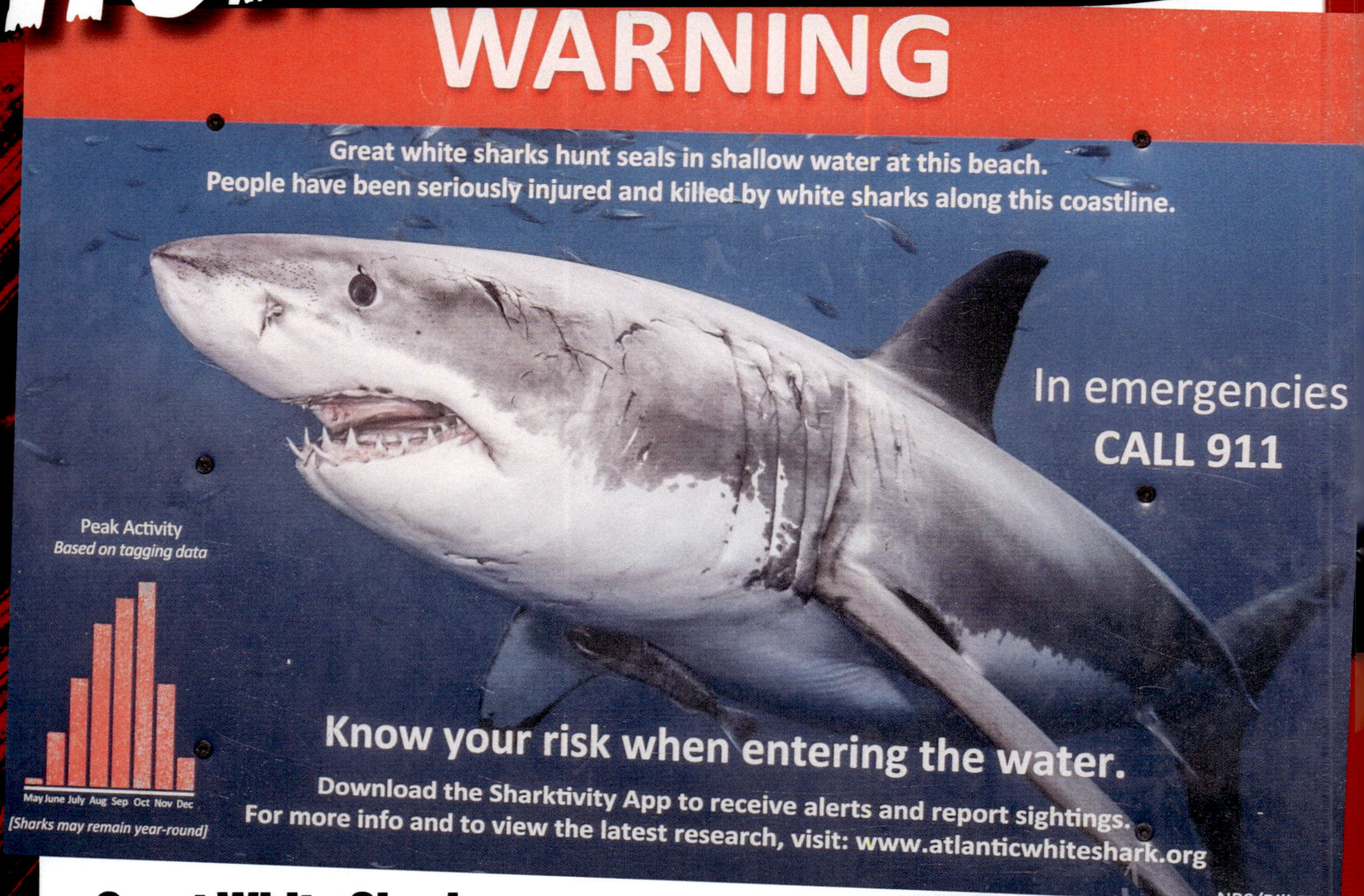

NPS/Bill Fisher

Great White Shark vs Human

Many people are terrified of great white sharks, but sharks rarely attack humans. Scientists estimate there are between 70 and 100 total shark attacks on people by all species of sharks around the world each year. Only about five of those attacks cause death. Cows, on the other hand, kill about 22 people in the United States alone each year.

MISUNDERSTOOD

Many great white shark attacks likely occur by accident. A shark may attack a person on a surfboard because it mistakes the surfer for a tasty seal or sea turtle. Often the shark will take a bite of a person's arm or leg and then swim away when it realizes its mistake. But sometimes the shark will stay to finish the job, especially if it is disturbed in its habitat or hungry.

To bring it to the surface, this shark has been lured, or tricked, with a seal **decoy**.

Even perfect predators make mistakes sometimes—and a person on a surfboard can look a lot like a seal or sea lion from below.

Hunting Sharks

People kill about 100 million sharks—including many great white sharks—each year. It is illegal to catch and kill great white sharks in many parts of the world, but some sport fishers still hunt them as trophies. Some people eat shark meat. Others hunt sharks for their jaws, teeth, and fins. The fins are often cut off and used to make shark fin soup. The rest of the shark is thrown away.

AT RISK

Deadly Waters

Great white sharks are also threatened by other human activities. They are often bycatch in commercial fisheries. That means they are accidentally caught in the fishing nets and other gear used to catch tuna, cod, salmon, and other fish for people to eat. The sharks—as well as dolphins, sea turtles, and other animals—may be thrown back into the water but don't always survive.

Climate Change

Like most marine animals, great white sharks are affected by **climate change**. Scientists have observed that the average temperature of Earth is rising over time. As oceans heat up, sharks are forced to migrate to new areas that have the temperate waters they need. They may also be forced to follow prey animals migrating to new areas of the ocean. These migrations disrupt the balance of ecosystems in both the usual and new shark habitats.

This young great white shark was caught as bycatch by fishers.

No Sharks Allowed

People in some coastal communities set up nets or drum lines to prevent shark attacks on their beaches. Drum lines are traps used to attract and catch large sharks with baited hooks. Great white sharks often get caught in the traps or tangled in the nets and die. If the sharks survive, they may be freed and released back into other areas of the ocean.

Shark nets keep people safe, but put great white sharks in danger.

SAVE THE SHARKS

This scientist is taking a blood sample from a shark.

Losing Numbers

Scientists do not know exactly how many great white sharks are left in the world because it is difficult to find and keep track of these wide-ranging animals. They do know that their populations are decreasing, however. Today, great white sharks are considered vulnerable, or at high risk of becoming extinct in the wild.

Studying Sharks

Scientists study sharks, including great white sharks, in the ocean. They measure the sharks and take samples from them. They put tags on the sharks that allow the scientists to track their movements in the sea. The tags gather important data such as where the sharks swim, the depths that they dive to, and the light levels and temperature of the water.

Marine Parks

Knowing where and when sharks go and gather can help protect them from danger. For example, governments can turn important shark habitats into marine parks. Marine parks are protected areas of the ocean where commercial fishing and other harmful human activities are not allowed.

Scientists may fly **drones** over coastal waters to observe sharks.

Tagging great white sharks allows scientists to gather key data about their movements and behaviors.

FREE TO ROAM

Come to the Shark Side

Unlike many vulnerable species, great white sharks cannot be bred or confined by humans. Even massive tanks in zoos and aquariums are too small for these mighty fish. The sharks refuse to be fed in the tanks and often ram their heads into the walls until they die. Great white sharks need to swim long distances each day and hunt for their own food in the ocean.

PREDATOR POP QUIZ

Imagine you are planning a cage diving trip to see great white sharks. Make a list of things you know about these sharks. Which ocean would you visit? What time of day would you enter the water? How would you identify a great white shark? Would they be swimming alone or with other sharks? What prey animals would the shark be hunting? Add other information you have learned about these perfect predators to your list.

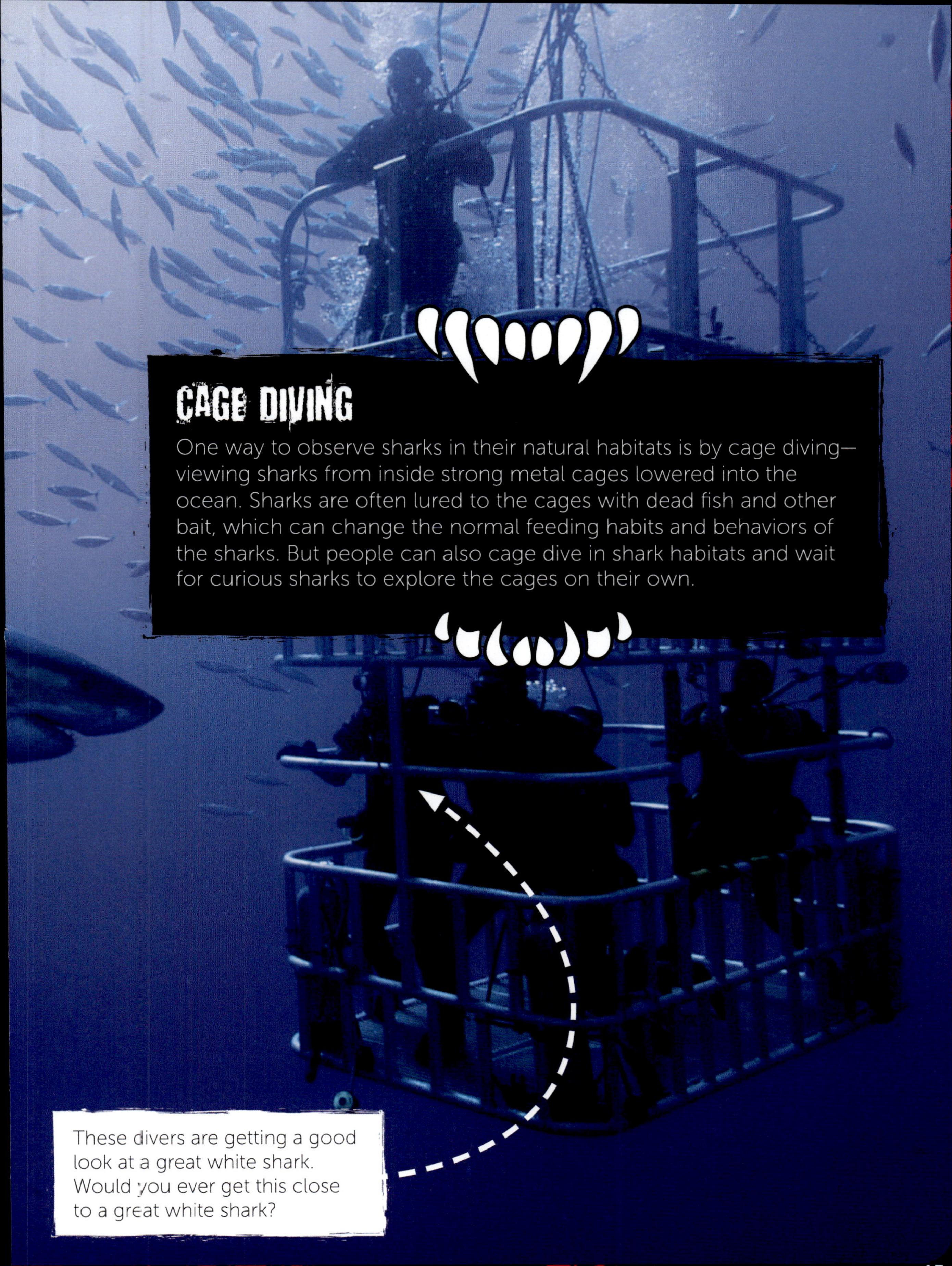

CAGE DIVING

One way to observe sharks in their natural habitats is by cage diving—viewing sharks from inside strong metal cages lowered into the ocean. Sharks are often lured to the cages with dead fish and other bait, which can change the normal feeding habits and behaviors of the sharks. But people can also cage dive in shark habitats and wait for curious sharks to explore the cages on their own.

These divers are getting a good look at a great white shark. Would you ever get this close to a great white shark?

GLOSSARY

adapted (uh-DAPT-ed): Changed over time to better suit one's environment

apex predators (EY-peks PRED-uh-ters): Predators with no natural predators of their own

breed (breed): Mate to have babies

camouflage (KAM-uh-flahzh): Colors and patterns that help an animal blend in with its natural surroundings

climate change (KLAHY-mit cheynj): A change in the usual long-term weather patterns in an area. Global warming is the current period of climate change.

decoy (DEE-koi): An object or person used as a trick; an animal decoy is used to attract another animal

dorsal (DAWR-suhl): On or near the back

drones (drohns): Remotely operated aircaft

echlocation (ek-oh-loh-KEY-shuhn): The process by which animals navigate by making sounds and listening to echoes

ecosystems (EE-koh-sis-tuhms): Communities of living and nonliving things in their environments

extinct (ik-STINGKT): No longer existing on Earth

food chain (food cheyn): A network of living things that shows the order in which animals depend on each other for food

gills (gils): Organs used by some animals to get oxygen from water

lateral (LAT-er-uhl): Related to the side or sides

migrate (MAHY-greyt): Move from place to place, often with the seasons

opportunistic (op-er-too-NIS-tik): Feeding on whatever food is available

organs (AWR-guhns): Body parts with specific purposes

oxygen (OK-si-juhn): A colorless and odorless gas that plants and animals need to live

pectoral (PEK-ter-uhl): Related to the chest

populations (pop-yuh-LEY-shuhns): Total number of a species living in a certain area

pores (pawrs): Small openings in the skin

precision (pri-SIZH-uhn): Being accurate or exact

predators (PRED-uh-ters): Animals that hunt other animals for food

prey (prey): Animals that are hunted and eaten by other animals

stealth (stelth): Quiet and secretive movements

streamlined (STREEM-lahynd): Designed to move easily and smoothly through water or air

temperate (TEM-per-it): Not very hot or cold

womb (woom): The organ in some female animals in which babies develop; also called the uterus

INDEX

COMPREHENSION QUESTIONS

1. **Humans kill about ________ sharks each year.**
   a) 100 million
   b) 75 million
   c) 50 million

2. **True or False:** Great white sharks will hunt all animals—big or small.

3. **True or False:** Great white sharks are at high risk of becoming extinct.

4. **The lateral line that runs down a great white shark's body allows it to ________________ .**
   a) detect faint electrical signals
   b) swim, bend, and turn quickly
   c) sense movements and pressure changes in the water

5. **True or False:** Great white sharks are much bigger and heavier than orcas.

6. **A great white shark can reach up to 21 feet (6.4 m) and weigh _______ .**
   a) 5,000 pounds (2,268 kg)
   b) 4,500 pounds (2,040 kg)
   c) 5,500 pounds (2,495 kg)

7. **True or False:** Most great white shark pups survive their first year.

8. **Most scientists use the following name for great white sharks.**
   a) white pointers
   b) white death
   c) white sharks

1. c, 2. False, 3. True, 4. c, 5. False, 6. b, 7. False, 8. c

About the Author

Robin Johnson is a Canadian author and editor who has written more than 100 nonfiction books for children—everything from Animal Celebrities to Food Scientists in Action. When she's not writing, Robin enjoys ambushing her husband and two sons in strategy games and is always on the hunt for the perfect cupcake.

GREAT WHITE SHARKS ON THE HUNT

Author: Robin Johnson
Series Concept and Development: Janine Deschenes, Melissa Boyce, Ellen Rodger
Editors: Janine Deschenes and Melissa Boyce
Series and Book Design: Margaret Salter
Proofreader: Ellen Rodger
Print Manager: Candice Campbell
Illustrations: Margaret Salter: page 39 (inset seal and surfer)
Photographs: Amy Lutz / Shutterstock.com: page 38 (top); Emmanuel R Lacoste / Shutterstock.com: page 42 (top); All other images from Shutterstock.com

Crabtree Publishing

crabtreebooks.com 800-387-7650

In Canada: We acknowledge the financial support of the Government of Canada through the Canada Book Fund for our publishing activities.

Hardcover 978-1-0398-8087-0
Paperback 978-1-0398-8448-9
Ebook (pdf) 978-1-0398-8208-9
Epub 978-1-0398-8328-4

Published in Canada
Crabtree Publishing
616 Welland Avenue
St. Catharines, Ontario
L2M 5V6

Published in the United States
Crabtree Publishing
347 Fifth Avenue
Suite 1402-145
New York, NY 10016

Library and Archives Canada Cataloguing in Publication
Available at the Library and Archives Canada

Library of Congress Cataloging-in-Publication Data
Available at the Library of Congress

Printed in the U.S.A./CP2025